ISBN 978-0-260-97420-4
PIBN 11110362

1 MONTH OF
FREE
READING

at
www.ForgottenBooks.com

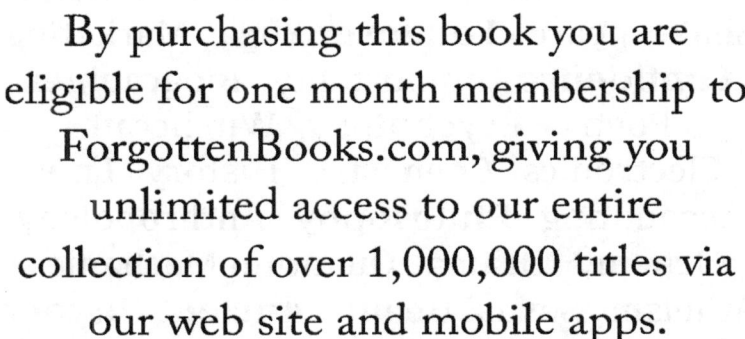

By purchasing this book you are
eligible for one month membership to
ForgottenBooks.com, giving you
unlimited access to our entire
collection of over 1,000,000 titles via
our web site and mobile apps.

To claim your free month visit:
www.forgottenbooks.com/free1110362

Historic, archived document

Do not assume content reflects current
scientific knowledge, policies, or practices.

BLAND APPLE SIRUP[1]

H. H. Mottern and R. H. Morris 3rd[2]

Eastern Regional Research Laboratory
Philadelphia, Pennsylvania

Bureau of Agricultural and Industrial Chemistry
Agricultural Research Administration
United States Department of Agriculture

Bland apple sirup may become an important item of commerce. Its wide acceptance as a humectant by cigarette manufacturers, the interest shown by other potential users, the manufacture of 3,000,000 pounds in the United States and Canada during the first year of its production, and an expected larger production from the 1943 crop indicate this possibility. Since the preliminary description of the process in August 1942, in mimeograph ACE-180, considerable additional information has been obtained on the process and on the properties and uses of the sirup. It was thought, therefore, that a new publication on this subject would be timely and useful.

Investigations on apple juice were undertaken with the view of eliminating the substances that impart flavor, odor, color, and jellification, leaving a more or less flavorless sugar solution that could be evaporated under vacuum to a bland sirup somewhat similar to commercial invert sirup, which is well known in the bakery, ice cream, soft drink, and other food industries. This objective has been achieved. Several industrial users have pronounced the sirup satisfactory, and a number of people have declared it to be an excellent table and cooking sirup.

COMPOSITION

Apple sirup made by the present process is amber in color, very sweet, and bland. It has no distinctive flavor, not even that of apple, but it has a slight bitter aftertaste, which is probably due to calcium malate. Its consistency is about that of an invert sugar sirup of the same solids content. At 75° F. its viscosity ranges from 800 to 1100 centipoises, with an average of about 1000.

The composition of apple sirup is given in Table I. A total soluble-solids content of 75 percent is given here not because that is a fixed value but because experience has shown that it is a satisfactory medium between 72 percent, the safe minimum for nonspoilage by micro-organisms, and 80 percent or above, at which point the sirup becomes molasseslike in consistency.

The value for total solids obtained by a Brix hydrometer spindle is higher than that obtained by the refractometer, as illustrated by the figures 77.6 and 75 percent in Table I. Since the refractometer is the more accurate for use with plant materials, it is used as the standard here. A sirup may, of

---

1/ Supersedes Mimeograph Circular ACE-180, PRODUCTION OF A BLAND SIRUP FROM APPLES, by M. A. Bradshaw and H. H. Mottern

2/ The authors are indebted to E. G. Beinhart, J. S. Hudnut, N. J. Gilbride, and C. S. Nevin for contributions to this paper

course, be sold on either basis.

A mixture of three sugars - levulose, dextrose, and sucrose - constitutes from 83 to 89 percent of the total solids, and hence about 65 percent of the sirup, as indicated in Table I. Although these sugars are equally valuable as food, the levulose is by far the sweetest and has the most pronounced property of absorbing and retaining moisture. Because of the high levulose content, there-fore, apple sirup is extremely sweet and, being moderately hygroscopic, can be used as a humectant. Since special interest thus attaches to the levulose content of apple juice, available data on this subject are given in Table II. The levulose is here calculated on the basis of the total solids and averages 50 percent; on the sirup basis this would be 37 percent, as indicated in Table I. The combination of the three sugars makes a desirable sirup, not only because of the sweetening and humectant properties resulting from the high levulose content but also because no one sugar will crystallize when the sirup dries. This means that even with low relative humidity some of the plasti-cizing effect of the sirup is retained.

The nonsugar solids of apple juice consist of small amounts of malic acid, tannin, pectin. soluble salts, coloring matter, and nitrogenous substances consisting of enzymes, proteins, and possibly some free amino acids.

TABLE I

Composition of Apple Sirup

		Percent
Total solids (as determined by refractometer)		75
Total solids (as determined by spindle)		77.6
Sugars		
Levulose	37 percent	
Dextrose	14 percent	
Sucrose	14 percent	
Total		65
Nonsugar solids		10
Ash		2.59
Calcium oxide		0.48

TABLE II

Levulose Content of Apple Products

		Levulose in total solids, %
Sirup, commercial		56
Sirup, commercial		57
Sirup, commercial		58
Concentrate, Stayman		50
Juice, clarified		51
Sirup from the above		48
Sirup, laboratory		45
Sirup, laboratory		45
Sirup, laboratory		42
Juice, Delicious [1]		50
Juice, 15 varieties [2]	minimum	45
	maximum	56
	average	47
Pulp, average of 11 samples [3]		55

[1] Bur. Standards Jour. of Research (1932)
[2] Ind. Eng. Chem. 9, 587 (1917)
[3] Plant Physiol. 17, 435 (1942)

FLOW SHEET FOR MANUFACTURE OF APPLE SIRUP

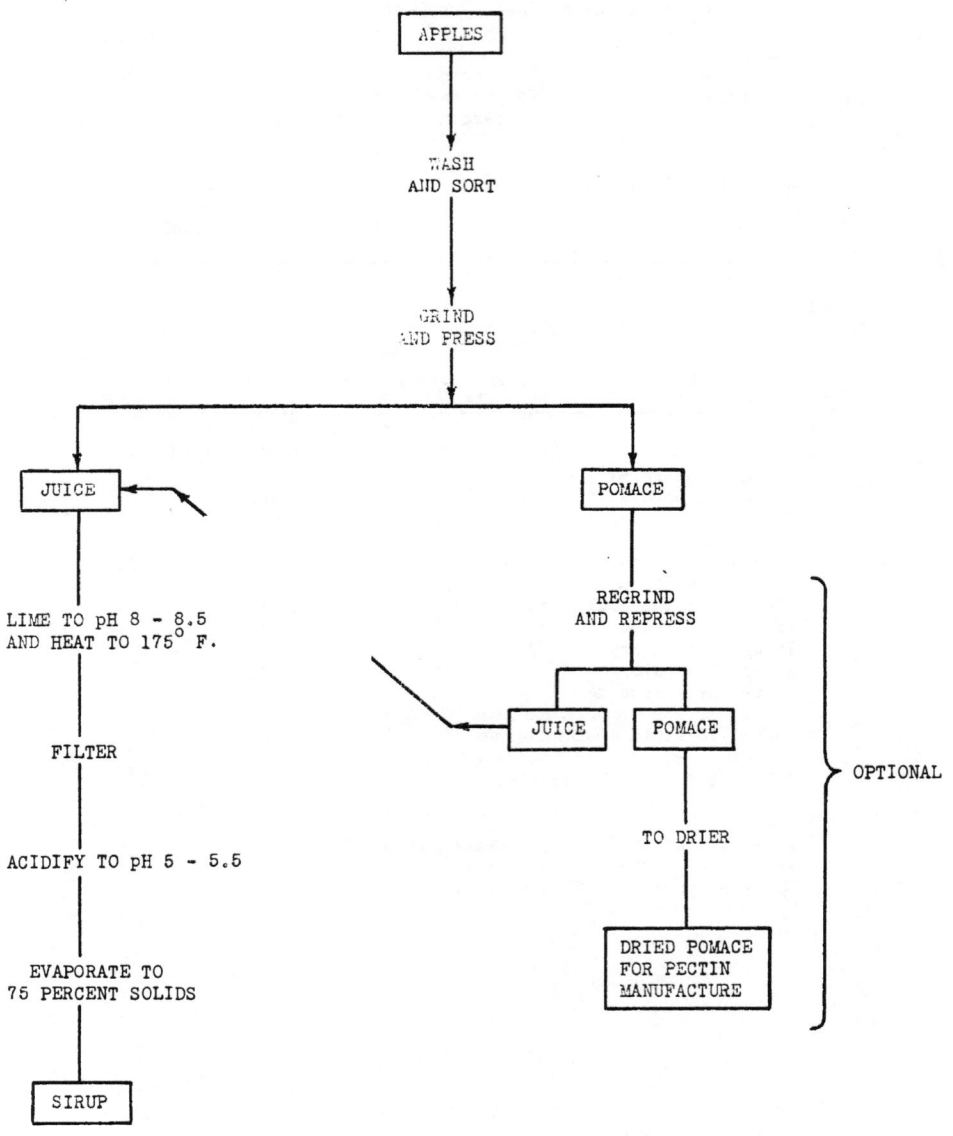

EQUIPMENT REQUIRED

The main items of equipment needed for the production of 6000 pounds of sirup per 24-hour day by the process described herein are as follows:

Sorting belt   )
Fruit washer   ) Not needed if juice
Hammer mill or grater)   is purchased
Hydraulic cider press)

Wooden tanks for storing juice
Small tank (25 gallons), or barrel, for preparing milk of lime
Two wooden tanks (2000 to 3000 gallons), with heating coils and stirrers, for liming the juice
Precoating tank, 200 to 300 gallons, with agitator
Two wooden tanks (2000 to 3000 gallons), equipped for agitation by compressed air or stirrer, for acidifying filtered juice
Sirup-receiving tank, 500 gallons, or larger if blending is desired
Filter press, plate-and-frame type, 200 to 300 square feet of filtering area, with pump
Air compressor for air stirring, if the latter is used
Vacuum pan, or other vacuum evaporating equipment, with necessary condenser, source of cooling water, and means of producing a vacuum
Boiler, 150 HP
Transfer pumps (one portable) for delivery of juice to storage and liming tanks
Sirup-withdrawal pump
Scale for weighing barrels
Barrels for finished sirup, oak, paraffin lined
Glass electrode pH meter, or colorimetric pH outfit
Refractometer, or Brix or Baume spindles

PROCESS

The process in general is illustrated in the flow sheet (page 4). It can be briefly outlined as follows: The apples are sorted, washed, and ground, and the juice is pressed out by a hydraulic press. The juice is treated with a slurry of hydrated lime until the pH is 8.0 to 8.5, heated to 175° F. (79° C.) to precipitate the pectin, and filtered. The clarified juice from the filter press is acidified with 1:3 sulfuric acid (or other acid) to a pH of 5.0 to 5.5 and then evaporated under vacuum to a sirup containing approximately 75 percent of solids.

Kind of Apples, Grading, Sorting, and Washing

Apparently any variety of apple may be used for sirup manufacture. Since the yield depends directly on solids content, varieties yielding juice of high solids content are more desirable. High acidity (0.6 percent or higher) may lead to precipitation of calcium malate, which would necessitate extra treatment to eliminate it. Summer varieties may be used, although they are inclined to be low in sugar and high in acidity.

Federal laws prohibit the use of unsound apples in food products, and hence preliminary sorting to remove unsound fruit is necessary. Juice from cider-grade fruit, sound but perhaps off-grade because of color, size, shape, or surface defects, is satisfactory. Immature fruit is to be avoided because it has a low sugar content. Over-ripe or frozen fruit is difficult to handle in a hydraulic press, but this difficulty can be overcome by the treatment indicated under the section "Grinding and Pressing."

Apples may also contain appreciable quantities of spray residues, which find their way into the juice upon extraction. In processing, the juice is reduced in volume by evaporation, and thus the spray residues are also concentrated. Therefore, to prevent excessive amounts of the residues in the finished sirup, it is advisable to wash contaminated fruit. Any of the accepted methods of washing for removal of spray residues may be used.

Cores and peels may be used for sirup designed for industrial uses. The yield of juice is about 135 gallons per ton, as compared with about 160 gallons for whole fruit. Since cores and peels contain most of the spray residue of the fruit, juice obtained from them is more likely to be contaminated. (See section headed "Contamination.")

Grinding and Pressing

The clean, sorted apples from the washer are passed by means of an elevator to a hammer mill or grater, where they are ground. The pulp is ground finer by the hammer mill and yields more juice, but the juice contains more fine particles of apple pulp and requires more filter aid for filtration. The grater requires less power, and this method is more adaptable to small installations.

The wet pulp is pressed by a hydraulic press. Soft, overripe fruit or fruit that has been frozen is difficult to press. Pressing such fruit is facilitated by mixing from 1 to 2 percent of a coarse diatomaceous filter aid with the wet pomace as the cheese is built up. Pomace from this fruit is not satisfactory for pectin manufacture because of its low pectin content.

If the pomace is to be dried, pressing it a second time has certain advantages. One is that from 10 to 15 gallons of additional juice per ton of fruit is obtained, and this will usually pay for the cost of the second pressing. Another is that this juice represents from 75 to 115 pounds of water that will not have to be removed in drying the pomace. A third factor is the higher pectin content of the dried pomace. Addition of water before the second pressing will result in additional extraction of sugar. Such juice is usually too dilute for vinegar making, but is satisfactory for sirup manufacture.

The advisability of drying pomace will naturally depend on the market for it. At present the demand is considerable. The price of pomace from the 1942 crop ranged from $45 to $90 per ton, and there are indications that it may be higher for the 1943 crop.

Juice Preservation

Storing preserved juice in wooden tanks makes possible an extension of the manufacturing season and thus reduces the overhead. The preservative should be added immediately after pressing.

Juice preserved by sodium benzoate at a concentration of 0.1 percent by weight of juice (14 ounces per 100 gallons) has been found satisfactory for sirup designed for tobacco uses. Sulfur dioxide at a concentration of 500 to 1000 parts per million preserves the juice satisfactorily, but it must be eliminated before lime is added. Otherwise the reaction of the lime with sulfurous acid will fix the sulfur dioxide in the juice in a stable form. Sulfur dioxide can be eliminated by heating the juice in the liming tank and blowing a stream of air into it through a cross or circle of copper tubing lying on the bottom of the tank. The copper tube should be provided with holes from three thirty-seconds to one-eighth inch in diameter drilled 2 or 3 inches apart. Provisions must be made to exhaust the fumes outside the plant.

Liming

The juice for processing is delivered from the press or storage tanks to the liming tanks, which are equipped with heating coils and agitators. The liming tanks should have a capacity of 2000 to 3000 gallons each and should be arranged for alternate use. This arrangement allows more time for heating, with less of a peak load on the boiler. The agitator should be an upright paddle type with sufficient speed to keep filter aid in suspension. The heating coils should clear the floor of the tank sufficiently to facilitate cleaning. As mentioned above, if sulfur dioxide has been added to the juice, it must be removed before liming.

From 2 to 3 pounds of lime will be required for every 100 gallons of juice. The lime should be of chemical grade, high in calcium oxide and low in magnesium and iron. As a rule, heating is started before the lime is added. Sufficient lime should be added to give a final pH value of 8.0 to 8.5. When the juice contains appreciable amounts of pectin, the pH can go as high as 9.0 to 9.5, since, when the reaction is complete, it will drop back to 8.0 to 8.5.

A slurry of lime is prepared by dispersing hydrated lime in water in the proportion of 1 pound of lime to a gallon of water. This is kept in a tank. It may be added by one of the following methods, but in all cases it should be thoroughly agitated while being added. The temperature of the juice may be brought up to 160°F. and all the lime slurry added at one time, a small stream being used to prevent localized action. Frequent pH readings must be made until the pH has reached a constant value of 8.0 to 8.5; then the juice is heated to 175° F. (79° C.). Another method is to allow the slurry to drain into the juice continuously during the heating period until the juice has reached a constant pH of 8.0 to 8.5. A third method embodies the combined principles of these two. An experienced operator can estimate the approximate quantity of lime that will be needed. When the juice has reached a temperature of 120° F. three-quarters of the calculated amount of lime is slowly added, and the pH is checked. Then at 140°, 160°, and 175° F. just enough lime slurry is added to raise the pH value to 8.0 to 8.5. An interval between each addition of lime is allowed to give the lime time to react completely with the pectin. Since the finished sirup may be darkened by treatment at a pH value above 8.0 at high temperatures, the pH value is never permitted to exceed the danger point.

A pH meter, preferably a glass-electrode type equipped with extension leads, should be used to control the liming. If such an instrument is not available, phenolphthalein indicator solution may be used. The latter requires more time, and the technique must be perfected to make a satisfactory determination. A funnel provided with a fast filter paper is supported above a beaker on a white background. Approximately 500 ml. of juice is taken from the tank, half this volume of coarse filter aid is mixed into it, and the mixture is poured into the funnel. After some of the clear juice has filtered through, a few drops of the indicator solution is added to it. A pink color indicates the end point of liming and corresponds to a pH value of approximately 8.25. If the test is applied to the cloudy unfiltered juice, the color will be hard to see, and the juice will reach a pH of 8.5 to 9.0 before a definite pink is noticed. If phenolphthalein is used, the intermediate pH values cannot be

taken; therefore, the best liming procedure would be to add approximately
three-quarters of the calculated amount of lime at 150° F. and then at 165° and
small quantities of slurry until the pink color appears. Care must be tak
to prevent adding an excess of lime, because the excess of acid required for
acidifying leads to danger of precipitating calcium salts during the concentration.

After the pH has been adjusted, filter aid is added, as indicated under
"Filtration."

After the juice has been brought to a final temperature of 175° F., it should
be held for 15 minutes to assure complete reaction of the lime with the pectin.
Owing to the risk of darkening at this high pH and temperature, the juice
should be filtered immediately after this holding period. Therefore the
filtration process must immediately follow the liming process.

Instead of coils in the liming tanks, a heat exchanger may be used. This con-
sists of tubes in a steam chest. As the cold juice is pumped through the
tubes, it is heated to 175° F. If the juice is heated over a long period of
time the load on the boiler is more evenly distributed. When the first tank
is full, the juice is turned into the second tank. The juice in the first
tank is then limed and filtered.

The use of a heat exchanger also affords the opportunity of using a continuous
process, which, however, should be planned by a qualified engineer. This set-
up requires two filter presses but only one small liming tank. This should be
equipped for agitation, and, to allow time for complete reaction, it should be
large enough to hold 15 minutes' supply of juice for the filter press. The
juice is passed through the heat exchanger and into the liming tank Here the
lime and filter aid are added continuously, so that the pH value remains at pH
8.0 to 8.5 and the concentration of filter aid does not fall below the calculated percentage. The lime and filter aid may also be injected directly into
the pipe line from the heat exchanger. When the first filter press has been
filled to capacity, the flow of juice is switched to the second press to allow
cleaning of the first. One of the acidifying tanks may also be omitted and
the acid added continuously, with continuous feed to the evaporator. Two pH
meters with extension electrodes would be required in order to insure complete
control of this process.

Filtration

At the end of the 15-minute holding period in the liming tank the juice should
be filtered. A plate-and-frame filter press, with a frame thickness of 1 3/4
to 2 inches and with bottom feed and top discharge, is most satisfactory, as
it accommodates a large quantity of sludge and can be easily dismantled for
removal of sludge and for cleaning and can also be quickly reassembled. The
large amount of sludge in the juice necessitates frequent cleaning of the
press. The tendency of the sludge to pack prevents its removal by backwashing.
Other types of filters, such as the Sweetland, may also be used. Wood and
cast iron have been found satisfactory materials for filters. The capacity of
a filter press averages about 5 gallons per hour per square foot of filtering
area but may vary widely from this, depending on the amount and nature of the
sludge in the juice.

Diatomaceous filter aid is used to filter the lime-treated juice. From 4 to 12
pounds per 100 gallons of juice (0.5 to 1.5 percent by weight) will be
required. In determining the proper amount, the maximum clarity with the
fastest rate of flow is used as the guide. Freshly pressed juice will require

more filter aid than juice which has been allowed to settle. Juice from firm apples is the easiest to filter. As the season progresses and the apples become softer, more filter aid must be used. If sufficient filter aid is not used, a slime is deposited on the cake. This packs tightly and causes the flow of juice to be stopped before the capacity of the press has been reached.

A precoat of filter aid approximately one-sixteenth to one-eighth of an inch thick should be applied to the filter cloths with either water or clarified juice. This will require 5 pounds of filter aid per 100 square feet of filtering surface. A filter aid of the same porosity as that used in the juice should generally be used. Since the precoat will do the initial filtering and be the base for the rest of the filtration, it should be applied properly. Before the filter aid is added to the precoating tank, water or clarified juice is circulated through the press until it is completely filled with the liquid. The precoating filter aid is then added to the liquid in the precoating tank with agitation, and the liquid in the press is rapidly recirculated to the precoating tank. When all the filter aid from the precoating tank has been applied to the cloths, juice from the liming tank is pumped through.

After the flow of limed juice has started, it should be reduced, without any cessation, to approximately the greatest rate consistent with good clarity. Every shift must be made promptly and efficiently so that there will be no interruption in the flow of juice, which would allow the press cake to fall or slough off. If water is used for precoating, the effluent must be recirculated to the precoating tank until the water is displaced. If juice is used for precoating, however, the precoating tank must be filled to the proper level with clarified juice for the next precoating operation.

Juice may be recirculated to the liming tank for a final check on clarity before going to the acidifying tank. When the effluent is sparkling clear, the juice is switched to the acidifying tank, but frequent checks are made on clarity to be sure that no leaks have developed in the press. The rate of flow may now be increased gradually over a period of 15 minutes until the maximum rate desired is obtained.

When the liming tank is empty, water should be run through the press to flush out the remaining juice until the effluent reaches approximately 2° Brix. This step is called sweetening off.

Reacidifying

The juice from the filter press is dark brown in color and sparkling clear. The pectin and other colloidal material have been removed, and the malic acid has been neutralized. The juice is now made slightly acid (pH value 5.0 to 5.5) in order to lighten the color and eliminate the alkaline taste. Sulfuric acid may be used if the sirup is intended for tobacco or industrial purposes. If intended for food, an edible organic acid should be used. Sulfuric acid should be diluted, before use, by adding one volume of U.S.P. grade concentrated sulfuric acid to three volumes of cold water, slowly and with stirring.

CAUTION! Never add the water to the acid. About one pint of dilute acid is required for 100 gallons of juice. The acid should be added slowly in a fine stream, and should be agitated to prevent localized action. A stream of air or mechanical paddles may be used for this purpose. During the acidification, the color of the juice changes from dark brown to a very light straw color.

Use of Activated Carbon

If a more highly clarified, very bland sirup is desired for certain purposes, the acidified juice may be treated with activated carbon in the proportion of 2 to 4 pounds of carbon per 100 gallons of juice, the amount depending on the grade of carbon and the degree of improvement desired. The carbon is added to the acidified juice, and the mixture is agitated to keep it in suspension. At the same time the temperature is brought to 150° F. and held for 15 minutes, and the juice is again filtered. Great care must be exercised to remove the carbon completely. A filter aid of fine porosity should be used, and the juice should be recirculated through the filter until it is certain that all particles of carbon are being retained. Instead of a filter aid, fiber filter pads of fine porosity may be used. On initial concentration, carbon-treated juice produces a lighter colored sirup, but it gradually darkens over a period of 2 weeks to nearly the color of the untreated sirup. For this reason, very little advantage can be attached to its use, and no commercial practice of using carbon has developed so far. Practically the same results may be obtained by using a filter aid of fine porosity in the initial filtration of the juice.

Evaporation

The clear, faintly acid juice is evaporated under vacuum to a sirup of about 75 percent solids. A vacuum of 26 inches or above is required, as high temperatures result in a dark sirup with a caramel flavor.

The temperature of the juice after filtration and acidification will usually range from 140° to 150° F. This is approximately the right temperature for it when drawn into the evaporator. The evaporator is filled to the desired level for operation, and thereafter the warm juice is introduced at the same rate that it is being evaporated. The operating level is kept constant until a concentration of 35 to 40 percent of solids is reached. Excess foaming during evaporation can be avoided by the addition of about two-thirds of an ounce of refined corn oil or other bland oil per 100 gallons of juice. When foaming becomes pronounced, usually at about 40 percent solids, and the evaporation rate declines, the batch should be finished off and a new one started. Having determined the holding capacity of the evaporator by a few trial runs, one can determine the most desirable size batch to evaporate and, in turn, the optimum size batch to lime and filter.

In testing the concentration of the sirup as it approaches the finishing point, a refractometer is preferable to a spindle, since the determination of solids content can be made quickly on a small sample and the instrument can be quickly cleaned for another determination. Accurate readings require a temperature correction, which may be made by means of the data in Table III. Provision should be made for withdrawing samples from the evaporator.

If a refractometer is not available, the solids content may be roughly but quickly determined by a Brix spindle. The reading of the spindle is greatly affected by temperature. The temperature must therefore be noted and a correction made by referring to Table IV.

A more accurate spindle determination may be made within 5 minutes by the following procedure: Carefully measure 230 to 250 ml. of sirup in a 500-ml. graduate. Add exactly the same volume of water as sirup and mix thoroughly. Then take a reading on this material with a spindle. After applying the temperature correction, this value may be converted to the soluble-solids value

TABLE III

Corrections for Refractometer Readings
at Temperatures Other Than 20° C.

Temperature of reading		For the approximate percentages of total solids indicated					
°C.	°F.	10	15	50	60	70	75

Subtract

15	59.0	0.27	0.31	0.36	0.37	0.36	0.36
16	60.8	.23	.26	.31	.32	.31	.29
17	62.6	.18	.20	.23	.23	.20	.17
18	64.4	.12	.14	.16	.15	.12	.09
19	66.2	.07	.08	.09	.08	.07	.05

Add

21	69.8	.07	.07	.07	.07	.07	.07
22	71.6	.14	.14	.15	.14	.14	.14
23	73.4	.20	.20	.23	.21	.22	.22
24	75.2	.26	.26	.30	.28	.29	.29
25	77.0	.32	.32	.38	.36	.36	.37
26	78.8	.39	.39	.46	.44	.43	.44
27	80.6	.46	.46	.55	.52	.50	.51
28	82.4	.53	.53	.63	.60	.57	.59
29	84.2	.60	.61	.71	.68	.65	.67
30	86.0	.67	.70	.80	.76	.73	.75

TABLE IV

Corrections for Brix Scale Readings
at Temperatures Other Than the Standard (68° F.)

Temperature of reading		For the approximate degrees Brix indicated					
°C.	°F.	10	15	50	60	70	75

					Subtract		
1.7	35	0.42	0.51	1.20	1.31	1.30	1.30
4.4	40	.41	.48	1.05	1.07	1.12	1.13
7.2	45	.39	.42	.87	.92	.93	.93
10.0	50	.35	.37	.68	.73	.73	.72
12.8	55	.28	.29	.50	.53	.53	.53
15.6	60	.20	.22	.32	.34	.33	.33
18.3	65	.10	.10	.12	.12	.12	.12

Add

21.1	70	.05	.05	.10	.08	.09	.08
23.9	75	.21	.23	.31	.30	.29	.28
26.7	80	.42	.44	.52	.50	.50	.48
29.4	85	.65	.68	.75	.71	.71	.70
32.2	90	.81	.84	.98	.93	.93	.92
	95	1.00	1.05	1.20	1.15	1.14	1.13
37.8	100	1.25	1.30	1.45	1.38	1.35	1.33
43.3	110	1.75	1.80	1.95	1.87	1.81	1.80
48.9	120	2.30	2.33	2.50	2.40	2.29	2.29
54.4	130	2.90	2.95	3.05	2.94	2.78	2.78
60.0	140	3.55	3.60	3.62	3.50	3.30	3.29
65.6	150	4.30	4.30	4.20	4.05	3.85	3.81
71.1	160	5.00	5.00	4.83	4.63	4.43	4.34
76.7	170	5.80	5.80	5.47	5.23	5.00	4.87
82.2	180	6.60	6.60	6.10	5.87	5.60	5.43

of the original sirup by means of the data in Table V.

A receiving tank may be provided to hold one or more batches of finished sirup. This permits quick emptying of the evaporator and the start of a new batch without delay. It also permits blending and allows the sirup to cool to slightly below the melting point of the paraffin coating of the barrels before being put into the barrels for shipment.

Three types of evaporators have been suggested for apple sirup, and operating conditions of each have been simulated in this Laboratory's experimental pilot-plant evaporator. The type generally employed for sirup manufacture as well as that usually available in the used-equipment market is a vacuum pan with steam jacket or heating coils or both. A second type is a thermal-circulation evaporator with an outside steam chest. This has a higher efficiency than the jacketed or coil type. A forced-circulation evaporator with outside steam chest has the highest efficiency. This type has a positive circulation rate, which insures a more positive surface contact as well as permits better control of the rate of evaporation.

Single-effect evaporation is considered to be more suitable for production of apple sirup than the multiple-effect type, as it entails less initial expenditure for equipment and operates at a lower temperature, reducing the risk of injury to the product. Because of the relatively limited quantity of juice available for processing at a given location and the short operating season, the savings resulting from multiple-effect evaporation are not sufficient to justify the greatly increased cost of the equipment.

It may be of help to give a single illustration of the relations between heating surface, vacuum, volume of condenser water, and amount of water evaporated. The following example, however, is merely illustrative. Engineering advice is usually required for individual cases.

A separator 5 feet in diameter with an outside steam chest having 350 square feet of effective heating surface (stainless-steel tubing) and using thermal circulation should have an evaporating capacity of approximately 6000 pounds of water per hour under a 27-inch vacuum. It is estimated that this would require approximately a 200-HP boiler (not including processing heat) using 4 to 6 pounds of coal per boiler HP per hour. Using a jet condenser with a barometric leg, this evaporating system would require approximately 330 gallons of cooling water per minute at 70° F.

Contamination

Copper has been used in commercial evaporators without excessive corrosion of the evaporator or significant contamination of the sirup. One sample from a commercial lot of sirup prepared in a copper vacuum pan from juice heated by copper coils and pumped through copper piping showed only 2.4 p.p.m. of copper. Copper equipment that has been idle must be thoroughly cleaned before use, as corrosion compounds may be readily dissolved by the sirup. For example, pear sirup prepared in previously idle copper equipment showed 73 p.p.m. of copper. In ordinary use of copper equipment, however, the contamination is insignificant, as shown by the results in Table VI.

Lead is removed in part by the liming and filtration treatments, and the quantity in the finished sirup may thus be reduced below the Federal tolerance. The amount of arsenic does not appear to be similarly reduced (Table VI) and may exceed the tolerance. The present tolerance (October 1943) is 0.050

TABLE V

Determination of Total Solids in Apple Sirup by Means
of Brix Spindle on Sirup Diluted 1:1 by Volume

Degrees Brix of diluted sirup (as determined by Brix spindle)	Total solids in original sirup (as determined by refractometer)
	%
32	56.3
33	57.7
34	59.2
35	60.7
36	62.2
37	63.6
38	65.1
39	66.5
40	67.9
41	69.3
42	70.8
43	72.2
44	73.7
45	75.2
46	76.6
47	78.0
48	79.5
49	80.9
50	82.4
51	83.8

TABLE VI

Metallic Residues in Apple Sirup

Sample No.	Description	Lead, grains per pound	Arsenic trioxide, grains per pound	Copper, p.p.m.
1	Laboratory sample prepared in stainless steel equipment	0.001	0.001	1.5
2	Commercial sample prepared in copper equipment with soldered joints	0.005	0.022	2.4
3	Commercial sample from peels and cores	0.031	0.055	
4	Sample 3, after passage through anion exchanger "A"		0.007	
5	Sample 3, after passage through anion exchanger "B"		0.047	
6	Pear sirup prepared in previously idle copper equipment	0.001	0.015	73

grain of lead per pound and 0.025 grain of arsenic trioxide (As_2O_3) per pound.

An effort has been made to remove arsenic from juice by passage through ion exchangers. A sirup containing 0.055 grain of As_2O_3 per pound was found after passage through an anion exchanger to contain 0.007 grain per pound. After passage through an exchanger made by another manufacturer, the sirup contained 0.047 grain of As_2O_3 per pound. The action of these exchangers in removing arsenic seems to be highly selective, and careful choice must be made if one is to obtain satisfactory results.

PRODUCTION RESULTS

A complete and accurate record of production data for each batch of sirup should be kept in a notebook. This is kept as much for future reference as for operation costs and records. A suggested form for such records is given in Table VII. A pint sample of each batch, with batch number and date of manufacture, should be held for reference purposes and examined for keeping qualities. The batch number should be marked on each barrel of sirup so that the sirup may be easily identified.

Type of juice includes such items as variety of fruit, whether the juice was obtained from whole fruit or peels and cores, and the method of preservation.

The form allows for adding the lime in stages. A constant pH of 8.0 should be reached in three or four stages, starting at approximately 140° F. The amount of lime used may be determined by making the slurry with a definite amount of lime per gallon and calculating the pounds from the number of gallons of slurry added. The column for remarks should contain any unusual fact concerning the liming procedure or other operations.

If water is used for precoating, the change from water to juice, when actual filtration has started, is too rapid for a Brix reading. When the press is sweetened off at the end, however, the process is gradual and should be stopped at some predetermined Brix value. A Brix of 2° would seem commercially feasible for this point. The amount of precoat should always be the same unless some leaves are cut out. Remarks on filtration and acidulation should contain such information as whether or not the juice was fairly clear or contained a lot of sludge. This will serve as a future guide for the amount of filter aid to be added to the juice. If the press fills and has to be cleaned, a note should be made and duplicate filter data set up for the second part. An unusually large or small amount of acid should be noted.

Data on evaporation should be recorded every hour, and intermediate changes recorded as made. Data on any other controls that the equipment has should be recorded under their respective subheadings.

The percentage yield is based on the theoretical amount of sirup of a given solids content that can be produced from a given amount of raw juice of determined solids content, if there is no loss of soluble solids. This theoretical yield is determined from Table VIII. For example, if 10,000 gallons of juice of 12.5 percent solids content are evaporated to a sirup of 75 percent solids content without loss, 14,670 pounds of sirup will be produced. Some losses are likely to occur, however, in the filter press operation, by entrainment and by spillage. For example, if 11,075 pounds of sirup are actually produced from the 10,000 gallons of juice, the yield will be 75 percent of theoretical.

TABLE VII

Production Data

Batch No._____ Date_____ Hour_____ Source_____ Type_____

Gallons_____ pH_____ Brix_____ Acidity_____ %

Heated from_____ °F. to_____ °F. in_____ hours_____ minutes at _____ # steam pressure

_____ lbs. of lime at _____ °F., pH_____)
_____ " " " " _____ " " _____)
_____ " " " " _____ " " _____) Remarks on liming
_____ " " " " _____ " " _____)
_____ " " " " _____ " " _____)

Precoated with _____ lbs. of _____   )
)
_____ lbs. of_____ added to juice.   )
) Remarks on
Filtration: Started_____ Finished_____ Sweetened off to Brix _____) filtration
) and acidu-
_____ hours_____ minutes to filter. Brix of batch____   ) lation
)
_____ quarts of 1:3 H_2SO_4, pH_____ to pH_____

Evaporation

		Juice		Condenser water			Pan		Steam pressure			
Time	Temp.	Gals. into pan	Vol. in pan	In	Out	Vac.	Temp. of material	Temp. of vapor	Boiler	Line	Heating surface	Remarks

Average_____

Rate of evaporation in gals./hr._____ Evaporator emptied, date_____ hour_____

Operator_____

Sirup:

_____ lbs. at_____ Brix._____ hours for complete batch. Yield_____ %

Remarks on color, taste, sediment, clarity, etc., of sirup.

TABLE VIII

Sirup Obtainable from 1000 Gallons of Juice
of Various Solids Contents

Solids in juice, %	Pounds of sirup with percentage of solids indicated							
	70	71	72	73	74	75	76	77
8.0	987	971	957	947	933	920	907	895
9.0	1115	1098	1083	1068	1053	1040	1027	1012
9.5	1172	1155	1138	1123	1107	1093	1079	1064
10.0	1243	1225	1207	1190	1176	1163	1145	1129
10.5	1300	1282	1262	1245	1229	1213	1198	1181
11.0	1371	1353	1332	1313	1296	1274	1263	1245
11.5	1429	1407	1386	1368	1351	1335	1316	1298
12.0	1500	1480	1458	1439	1420	1404	1384	1365
12.5	1571	1551	1526	1506	1486	1467	1449	1430
13.0	1630	1605	1582	1562	1540	1521	1501	1481
13.5	1701	1675	1651	1630	1605	1588	1566	1544
14.0	1763	1732	1708	1685	1660	1642	1619	1601
14.5	1830	1803	1775	1753	1728	1705	1685	1662
15.0	1900	1873	1844	1822	1795	1773	1750	1728
16.0	2023	1999	1972	1945	1917	1891	1870	1844

COST OF MANUFACTURE

Since there is generally a wide variation in conditions such as capacity, other products manufactured, availability and cost of apples or juice, and location, definite figures on costs would not only be valueless but even misleading. It is believed, however, that if an example of a cost estimate is given in which the assumptions as to plant set-up and other conditions are specifically defined, it would be useful as a guide. Therefore, a cost estimate is given below based on a plant operating under the following assumed conditions.

(a) Plant producing only apple sirup. A plant producing only apple sirup would have higher overhead costs than plants manufacturing other products in addition to the apple sirup. In the latter case, part of the overhead would be borne by the other products.

(b) Production of 6000 pounds of sirup daily for a 100-day season, a total of 600,000 pounds of sirup. This assumes a supply of approximately 5000 gallons of juice a day during the normal apple season of about 100 operating days. If the season is extended beyond the 100-day period, or a larger quantity is processed, the overhead cost per pound of sirup will be correspondingly reduced.

(c) Rent for building - $100 per month

(d) Cost of equipment - $30,000. Details of this estimate are given under "Plant Equipment".

(e) Juice of 12.5° Brix costing 12 cents a gallon and yielding 1.32 pounds of sirup per gallon with 10 percent loss of juice. It was decided to start with juice, rather than apples, in making this cost estimate, as in some cases the production of juice may be a business in itself and there might be some confusion as to what constitutes a profit on juice and what a profit on sirup. The assumption of 12.5° Brix juice costing 12 cents a gallon and yielding 1.32 pounds of sirup per gallon, with 10 percent loss of juice, is based on what is considered an average condition at present.

(f) Jet condenser

(g) Purchased water. If the unit cost of purchased water is high, consideration should be given to the possibility of either re-using the water after spray cooling, or of drilling wells.

SUMMARY OF OPERATING COSTS

	Cost per pound of sirup
Juice	$0.0910
Materials	.0068
Shipping containers	.0105
Fuel	.0023
Water	.0019
Labor	.0111
General overhead	.0400
Total	$0.1636

BREAKDOWN OF COSTS

Direct Operating Costs

Cost per pound of sirup

Juice, 12.5° Brix, at $0.12 per gallon, delivered $0.0910

Materials
    a. Preservative 0.1 percent sodium benzoate, 8.75 pounds
       per 1000 gallons of juice at $.50 per pound $0.0032
    b. Lime - 30 pounds per 1000 gallons at $20 per ton,
       delivered 0002
    c. Filter aid - 1 percent of weight of juice, 0.0875 pound
       per gallon at $92.20 per ton, delivered 0031
    d. Acid - H_2SO_4, 12 pounds per 1000 gallons at $0.0325 per
       pound 0003
                               $0.0068   .0068

Shipping containers - 50-gallon, paraffin-lined, oak barrels (holding
590 pounds of sirup) at $6.20 each, delivered 0105

Fuel - 0.790 pound of coal per pound of sirup at $6.00 per ton . . .0023

Water - 180 gallons per minute at $0.40 per 1,000 cubic feet or
$0.00005 per gallon of water, 302,400 gallons of water for 8,020
pounds of sirup, or 37.7 gallons of water per pound of sirup 0019

Labor	Hourly rate	Per day		
1 Engineer	$0.90	$ 7.20		
2 Firemen	.85	13.60		
3 Operators	.90	21.60		
5 Helpers	.60	24.00		
		$66.40	. . . .	.0111

General Overhead

Manager and plant superintendent $3,600
Office help (stenographer and bookkeeper) 1,560
Rent ($100 per month) 1,200
Supplies
    a. Office, postage, etc. 300
    b. Factory, laboratory 2,000
Telephone and telegraph 600
Power and lights 200
Travel and miscellaneous 1,000
Insurance, 2 percent of $30,000 plant 600
Compensation and Social Security deductions (0.355 percent of
    salaries and wages) 419
Maintenance and repairs (10 percent of $30,000-plant) . . . 3,000
Depreciation (10 percent of $30,000-plant) 3,000
Interest on investment (6 percent of $30,000-plant) 1,800
Operating capital (6 percent of $35,000) 2,100
Taxes (2 percent of $30,000-plant) (no profit tax) 600
Legal and accountant fees 200
Samples, containers, and express 300
Cartage 1,000
Miscellaneous 521
                                  $24,000 .0400

      Total $0.1636

PLANT EQUIPMENT

Evaporator and condenser	$12,000
Installation of evaporator and condenser	1,000
150-HP boiler (used)	1,000
Boiler accessories (pipes, valves, reducers, etc.)	1,500
Boiler installation (includes moving to plant)	1,500
Tanks	
Storage, seven 5000-gallon	3,500
Processing, with coils and agitators, two 2500-gallon	1,500
Acidifying, with agitators (air), two 2500-gallon	1,000
Sirup storage, 1500-gallon	400
Precoating, with agitator	200
Filter press, plate-and-frame, open delivery, with pump and motor	1,500
Pumps	
Raw-juice receiving)	
Raw juice to processing tank)	
Filtrate to acidifying tank)	
Sirup to storage tank)	1,000
Factory installation	2,000
Office equipment and miscellaneous	1,900
Total	$30,000

JUICE

	Cost per gallon
Apples - $0.60 per 100 pounds	$0.0750
Pressing, overall	.0250
Delivery cost	.0050
Profit (14.3 percent of $0.105)	.0150
Total	$0.1200

MARKETS

The outstanding characteristics of apple sirup are hygroscopicity and sweetness. As a hygroscopic agent, it compares favorably with glycerin, and it is approximately 30 percent sweeter than a sugar cane sirup of equal solids content. Wherever these characteristics are desirable, there is a potential market.

Because of its peculiar humectant and plasticizing properties, apple sirup was immediately accepted when it was first offered to the tobacco industry, in May 1942. At that time, supplies of glycerine, diethylene glycol, propylene glycol, and sugars, all used in rather substantial quantities as humectants in tobacco processing, were curtailed. This focused attention upon the merits of apple sirup, and by May 1943 there was a demand for no less than 20 million pounds of the sirup for use in tobacco products. It is believed that apple sirup will find a permanent place in the tobacco industry.

Because of its high levulose content, apple sirup is a satisfactory substitute for glycerin in tobacco products. It has about 80 to 85 percent of the hygroscopic value of glycerin, and it has the further advantage of being nontoxic in itself and not imparting a foreign taste to the product. Its products of combustion are also nontoxic and nonirritating.

In preparing tobacco for the cutting machine, the mass of leaf must be of suitable pliability to allow the use of high-speed cutting knives and to reduce to the minimum the production of dust and short pieces in the cutting operation. Apple sirup supplies this desirable plasticizing effect. It is applied either by dipping the tobacco in a solution of the sirup or spraying the sirup on the mass of tobacco.

Apple sirup blends exceptionally well with Burley tobacco, and evidence is accumulating that it favorably affects the aromatics of the air-cured types of tobacco. It is with these types that the long-time use of apple sirup may be expected.

In addition to its use in the tobacco industry as a humectant and sweetener, apple sirup is finding application in products such as ice cream, sherbets, ices, pastry, candy, table sirups, cork gaskets, paper, textile processing and finishing, pharmaceuticals, cosmetics, and sugar curing of hams. It also has interesting possibilities as a modifier for milk because of its effect in lowering the curd tension.

By modifying the process, it is possible to prepare apple sirup with special properties to meet the requirements of particular markets. For instance, the calcium should be removed when the sirup is to be used in a product containing soap. This can be accomplished by the use of ion exchangers. If a more viscous solution is desired, the natural pectin in the fresh juice is not removed. Sirups with these modifications have been prepared on a laboratory scale.

CPSIA information can be obtained
at www.ICGtesting.com
Printed in the USA
LVHW011822061118
596180LV00013BA/1007/P